EDI[illegible]ARIO
MUSHROOMS

A guide to common species

Jeffrey C. Domm
with Ben Kendrick

Formac Publishing Company Limited
Halifax

I would like to dedicate this book to all the memorable outdoor adventures camping with my siblings, Sue, Mark and Bob. —JD

Formac Publishing Company Limited recognizes the support of the Province of Nova Scotia. We are pleased to work in partnership with the Department of Communities, Culture, Tourism and Heritage to develop and promote our cultural resources for all Nova Scotians. We acknowledge the support of the Canada Council for the Arts.

NOVA SCOTIA Canada Canada Council for the Arts Conseil des Arts du Canada

The information in this book is intended to serve as a resource. Anyone who forages for and chooses to consume wild mushrooms does so at their own risk. Decisions to eat mushrooms included in this book need to be taken with great caution, care and absolute certainty. The authors and the publisher are not liable for any consequences of consuming the mushrooms included in this book.

Cover design: Tyler Cleroux
Cover image: Jeffrey C. Domm
All images Jeffrey C. Domm except p. 20 (iStock); p. 21B (Alamy)

Library and Archives Canada Cataloguing in Publication

Title: Edible Ontario mushrooms : a guide to common species / Jeffrey C. Domm, with Ben Kendrick.
Names: Domm, Jeffrey C., 1958- author | Kendrick, Bengi, author.
Description: Includes index.
Identifiers: Canadiana 20250314363 | ISBN 9781459508248 (softcover)
Subjects: LCSH: Edible mushrooms—Ontario—Identification. | LCSH: Edible mushrooms—Ontario. | LCGFT: Field guides.
Classification: LCC QK605.7.O6 D66 2026 | DDC 579.6/163209713—dc23

Formac Publishing Company Limited
5502 Atlantic Street
Halifax, Nova Scotia, Canada
B3H 1G4
www.formac.ca

Printed and bound in Korea.

Contents

STOP! Read this before foraging

With over three thousand mushroom species in Atlantic Canada, the odds of misidentifying a toxic mushroom are high if you are not trained to distinguish between edible, inedible and poisonous mushrooms.

Foraging Rule #1: Never rely on a field guide to determine whether a wild mushroom is edible.

This field guide is an initial step in learning about common mushroom species in Ontario. With plenty of look-alike species, the difference between an edible mushroom and a poisonous one can be as subtle as a faded patch of colour that only an expert could discern.

For example, Field mushrooms (*Agaricus campestris*) look just like Destroying angels *(Amanita amerivirosa)*. One is a common edible mushroom featured in this book (page 38). The other can cause fatal liver and kidney damage. No field guide can guarantee whether a mushroom is edible or toxic.

Foraging Rule #2: Never eat a wild mushroom that an expert hasn't identified.

If you want to start harvesting wild mushrooms, consider joining an established group in your area. Organizations like the Mycological Society of Toronto, Ontario Mushroom Hunters and Foragers, Southern Ontario Mushroom Hunters, and the North American Mycological Association are all great resources. Wild Muskoka Botanicals hosts many foraging classes in the Muskoka area.

Foraging Rule #3: In case of an emergency, know who to call.

Always have the phone number for Poison Control handy, just in case. The Canada-wide, toll-free number is 1-844-POISON-X, or 1-844-764-7669.

Foraging Rule #4: Cook wild mushrooms before eating.

Wild mushrooms should never be consumed raw. Cooked mushrooms are less likely to cause gastrointestinal distress.

Foraging Rule #5: Never forget the old saying: "There are old mushroom hunters, and there are bold mushroom hunters, but there are no old, bold mushroom hunters."

Other tips for safe and sustainable foraging

1. **Transport your mushrooms in a mesh bag or basket:** This protects them from damage while allowing spores to disperse in the forest.

2. **Harvest gently:** The base of the stem can provide key identifying information, so harvest the whole mushroom by gently digging it up with your hands or a trowel instead of cutting the stem with a knife.

3. **Bring gloves:** The texture of a cap or stem may reveal essential clues that can only be gleaned by touching the specimen with your bare hands. However, contact with some mushroom species not featured in this book may cause adverse skin reactions, so show caution before touching and bring gloves as a precaution.

Visual Index

Identifying mushrooms by colour is a challenge. Colours vary widely within specimens of the same species, and the colour of a single specimen often changes over time. While this index shows some of the more diversely coloured mushrooms, and their page numbers, remember that colour is only one of the many characteristics to consider when identifying a specimen.

38

28

80

34

58

62

56

30

26

40

58

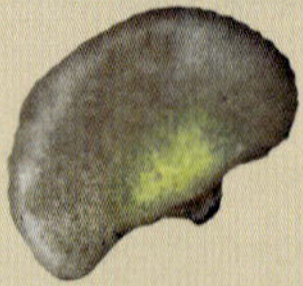
22

24

72

64
78
36
50
76
58
22
68
52
46

26
54
74
32
48
44
42
70

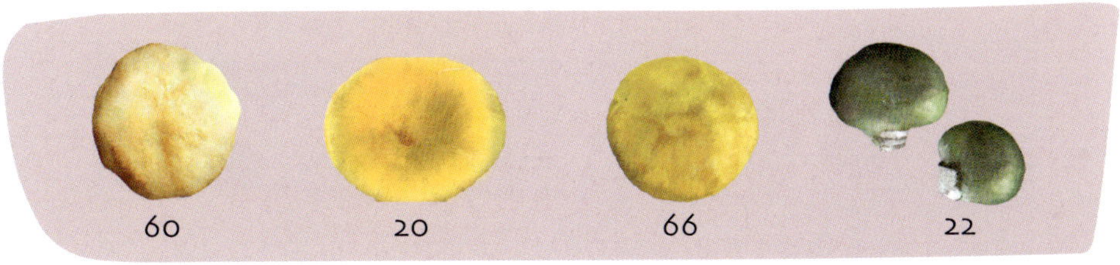
60
20
66
22

Seasonal Index

Common Name	Latin Name	Page	Availability
Yellow swamp brittlegill	*Russula claroflava*	20	J F M A M **J J A S O** N D
Late oyster	*Sarcomyxa serotina*	22	**J** F M A M J J A S O **N D**
Slimy spike	*Gomphidius glutinosus*	24	J F M A M J J A **S O N** D
Blewit	*Collybia nuda*	26	J F M A M J J A **S O N** D
Purple and white deceiver	*Laccaria ochropurpurea*	28	J F M A M J **J A S O** N D
Indigo milkcap	*Lactarius indigo*	30	J F M A M J **J A S O** N D
Brick cap mushroom	*Hypholoma sublateritium*	32	J F M A M J J A **S O N** D
Field mushroom	*Agaricus campestris*	34	J F M A M J **J A S O** N D
Dark honey mushroom	*Armillaria ostoyae*	36	J F M A M J J **A S O** N D
Aspen oyster mushroom	*Pleurotus populinus*	38	J F M A M **J J** A S O N D
Shaggy mane	*Coprinus comatus*	40	J F M A M J J A **S** O N D
Larch bolete	*Suillus grevillei*	42	J F M A M J **J A S O** N D
Smoothish-stemmed bolete	*Hemileccinum subglabripes*	44	J F M A M J **J A S** O N D
King bolete	*Boletus chippewaensis*	46	J F M A M J **J A S** O N D
Bay bolete	*Imleria badia*	48	J F M A M J J **A S O** N D

Common Name	Latin Name	Page	Availability
Dryad's saddle	*Cerioporus squamosus*	50	J F M A M J J A S O N D
Brown birch bolete	*Leccinum scabrum*	52	J F M A M J J A S O N D
Chicken-of-the-woods	*Laetiporus sulphureus*	54	J F M A M J J A S O N D
Bear's head tooth fungus	*Hericium americanum*	56	J F M A M J J A S O N D
Toothed jelly fungus	*Pseudohydnum gelatinosum*	58	J F M A M J J A S O N D
Wood hedgehog	*Hydnum repandum*	60	J F M A M J J A S O N D
White coral fungus	*Ramariopsis kunzei*	62	J F M A M J J A S O N D
Crown tipped coral	*Artomyces pyxidatus*	64	J F M A M J J A S O N D
Golden chanterelle	*Cantharellus enelensis*	66	J F M A M J J A S O N D
Winter chanterelle	*Craterellus tubaeformis*	68	J F M A M J J A S O N D
Cinnabar chanterelle	*Cantharellus cinnabarinus*	70	J F M A M J J A S O N D
Black trumpets	*Craterellus fallax*	72	J F M A M J J A S O N D
Lobster mushroom	*Hypomyces lactifluorum*	74	J F M A M J J A S O N D
Shrimp of the woods	*Entoloma abortivum*	76	J F M A M J J A S O N D
Common puffball	*Lycoperdon perlatum*	78	J F M A M J J A S O N D
Giant puffball	*Calvatia gigantea*	80	J F M A M J J A S O N D

What is a Mushroom?

Mushrooms are not plants. Rather, they are the "fruit" of fungus, a relationship similar to that of a flower to a flowering shrub.

Most stages in a mushroom's life cycle are completely invisible to us. However, a close examination of their growth reveals a complex and captivating spectacle of some of the most beautiful natural forms on Earth. Their vibrant colours and exotic shapes rival those of coral reefs beneath the sea. And it can all be observed right in your own backyard or along a path in a local park.

A mushroom's life cycle begins with the production and release of spores, usually from the gills or pores found under a mushroom's cap. Each mushroom can generate tens of thousands to millions of spores, which are dispersed through the air or the water to establish new growth elsewhere, much like the seeds of a plant. Spores are single-cell structures that germinate to produce hyphae. These hyphae then combine into vast, web-like networks called mycelia. Much like a plant's root system, mycelia will absorb nutrients, typically from the soil or rotting tree trunks, to support the growth of new fruiting bodies to continue the cycle.

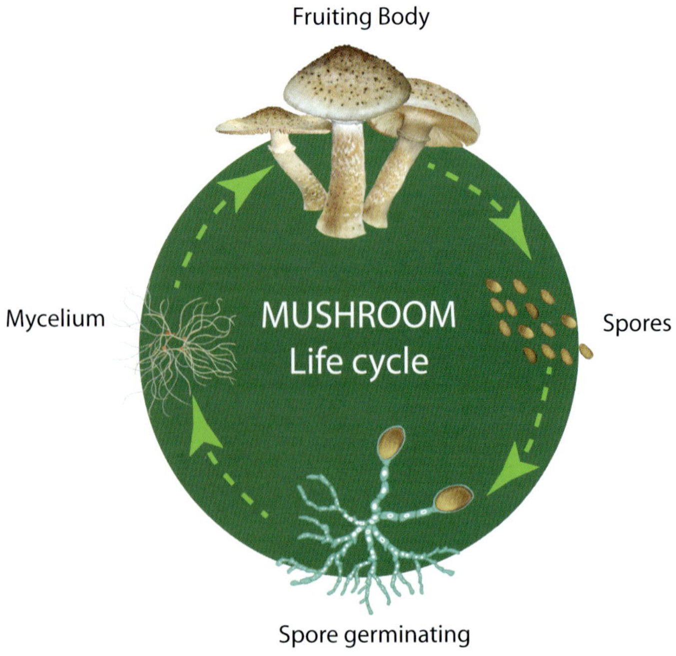

Mushroom Anatomy

Gilled Mushrooms (pages 20–41)

Gilled mushrooms have gills, paper-thin, plate-like structures hanging vertically below the cap.

If mushroom anatomy is new to you, a good place to start learning to identify a species properly is to look at a White button mushroom. Cut the mushroom in half from top to bottom using a sharp knife to expose a cross-section of the gills and inner flesh.

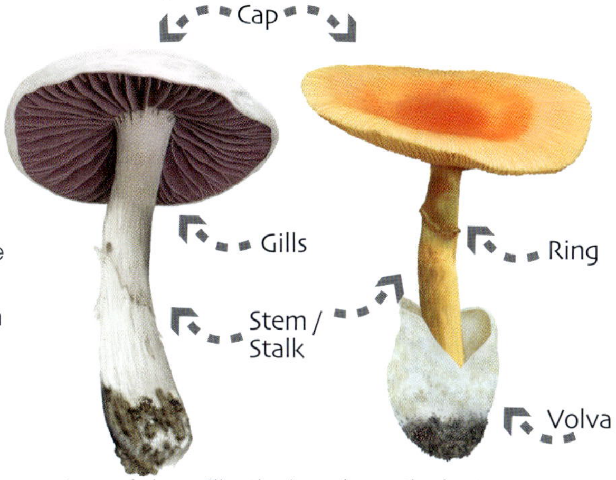

Look closely at the spacing of the gills, their colour, their structure and how they attach to the stalk. These features help identify the species of the gilled mushroom.

Poroid Mushrooms (pages 42–55)

Poroid mushrooms, also called boletes, have spongy pored undersides growing in place of gills or teeth. Their umbrella-shaped caps have a unique spore-producing feature consisting of hundreds, even thousands, of small holes formed by tubes inside the cap. Spores are produced inside these tubes and then released into the air for propagation.

Toothed Mushrooms (pages 56–61)

Mushrooms with teeth sound threatening, but these teeth describe the thin, spike-like structures that hang under the fruiting body's cap. The teeth, designed for spore production and dispersal, may be long or short and resemble stalactites hanging from the ceiling of a cave.

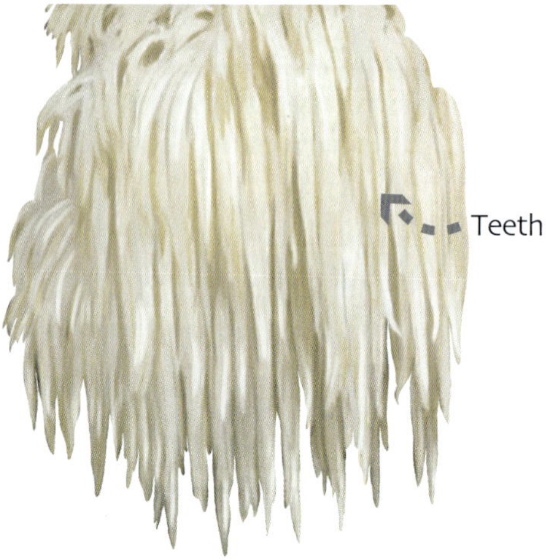

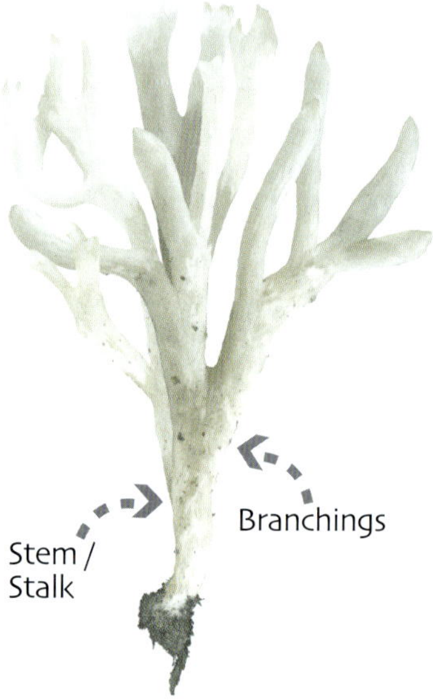

Coral Mushrooms (pages 62–65)

The name says it all! The fruiting bodies of these fungi are branch-like forms with stalks but no caps, stems or other typical identifying features. Instead of using gills or pores for propagation, coral mushrooms have branches covered with structures called basidia that produce spores.

Chanterelle-like Mushrooms (pages 66–73)

These funnel-shaped mushrooms often have wrinkle-like folds on the underside or stalk called "false gills."

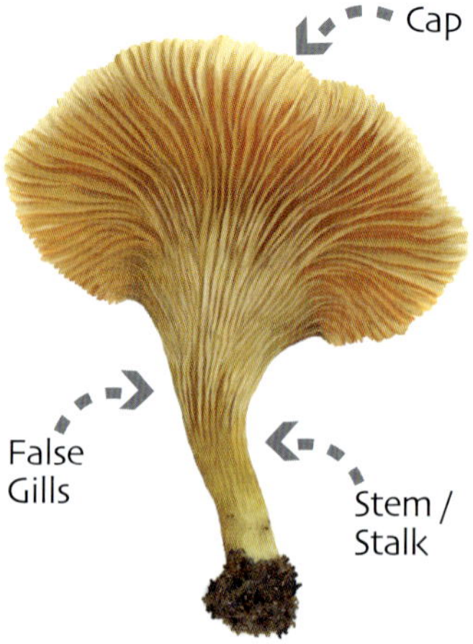

Parasitic Mushrooms (pages 74–77)

Parasitic mushrooms are the product of one species of fungus taking over another fungus or plant for its own nourishment. Depending on the specimen's age, you may be able to see traces of the host plant or mushroom before the parasitic species transforms the host beyond recognition.

Puffball Mushrooms (pages 78–81)

Puffball mushrooms are recognizable for their strange shapes and lack of a visible stem, though some may have a pseudo-stem.

How to Use this Book

Identifying mushrooms is a visual process that requires attention to shape, colour, size, location, habitat and season. This book provides highly realistic digital illustrations from a variety of angles to show the wide range of characteristics a single species can adopt. When identifying a mushroom, it is important to consider every characteristic, including smell, texture and bruising colour, to distinguish between similar species.

1. **Common and Latin names:** Common names will often refer to multiple species with very similar characteristics, so the Latin names are included to specify the precise species being described.

2. **Silhouette icon:** Solid outlines of each mushroom type are included to help quickly locate a specific species or section.

3. **"Look out for look-alikes" or "Good to know":** Here you will find important information about look-alike species or other useful information.

4 **Must-have features:** This list identifies 4–6 distinct features of each species.

5 **Habitat:** Mushrooms grow on the ground, on living and dead trees or on other mushrooms. For some mushrooms, habitat is a key identifying characteristic, while others will grow just about anywhere.

Decaying log

Deciduous trees

Coniferous trees

Mixed forest

Forest floor

Tree trunk

Grass

Near water

6 **Spore colour:** Unless there are spores that have accumulated on the skin of the mushroom, taking a spore print may be necessary to determine the colour of the spores. See page 20 for instructions.

7 **Bruising colour:** After harvesting, firmly grasp or pinch a part of the mushroom — bruising and bruise colour can be key identifying information.

8 **Smell:** A mushroom's smell, or its absence, is an important identifying feature.

9 **Growth season:** Mushrooms are most common in late summer and early fall, but some species appear in spring and even early winter. The calendar will let you know when a mushroom is most likely to appear.

Glossary

Agaric: A fungus that produces spores on gills.

Bolete: Fungi which have pores instead of gills.

Bracket: Poroid fungi with a shelf-like structure growing from trees or dead wood.

Cap: Umbrella-like structure on top of a mushroom that holds gills or pores.

Convex: Rounded or domed cap shape.

Crown tips: The ends of branches of coral mushrooms.

Flesh: The soft inner parts of a mushroom fruit body.

Fruit body or sporing body: The visible structure of fungi which supports and produces spores.

Fungus (or fungi if plural): A group of spore-producing organisms which feed on organic matter.

Gills: Linear plate-like structures found on the underside of a cap which produce and deliver spores for reproduction.

Habitat: The plant life, soil and any other distinctive components of the place where the fungus naturally occurs.

Hydnoid fungus: Fungi with tooth-like or spine-like spore-producing structures.

Jellies: Mushrooms that are transparent, jelly-like or rubbery in form.

Milk: A liquid released from certain mushroom species.

Mycelium: Microscopic fibres with a root-like structure that grow in soil, or in a host such as a tree or felled log, out of which the fruiting body of the mushroom grows.

Polypores: Fungi species with pores under their caps, also known as poroid mushrooms.

Pores: Small holes under the cap of a mushroom out of which spores are dispersed. Pores are the end of a series of small tubes in the cap.

Ring: A collar-like structure on the stalk that is leftover material of a partial veil rupturing to expose gills or other spore-producing structure.

Scabers: Small growths appearing on a stalk resembling new hair growing back after shaving.

Spores: Reproductive seed-like forms that are released through the mushrooms fruit body and dispersed into the air or water.

Spore print: Spores disbursed by mushroom cap when left for a length of time on paper. Can be a key identifier in mushroom identification.

Stalk (stem): A stipe, or stem that supports the cap of the mushroom.

Teeth: Spore-generating structures found under the cap of Toothed mushrooms.

Tubes: Tube-like structures found inside the cap of a poroid species, visible only if cut open.

Veil: A thin layer of tissue that covers and protects the fruit body of immature mushrooms.

Volva: A cup-shaped structure found at the base of some mushrooms that is the remnant of the universal veil.

DIY:
How to Take a Spore Print

Spores are microscopic, so taking a spore print renders the invisible visible. This simple experiment requires a bit of additional work but will offer you key identifying information.

Materials

- Two caps of the same mushroom
- One piece of white paper
- One piece of black paper
- Two cups or jars large enough to cover the specimens and leave additional space around them.

Step 1: Place your mushroom caps on either piece of paper, with gills or pores facing down.

Step 2: Using your finger, collect and let a drop of water fall onto each cap. This small amount of moisture will help maintain ideal spore-bearing conditions.

Step 3: Cover each mushroom with a glass or jar until a stain appears on the paper. This process can take between two and twenty-four hours, depending on humidity and the mushroom's freshness.

Step 4 (optional): If you want to preserve your spore print, spray with artist fixative or hairspray.

Note: To avoid damaging the mushroom by removing the cap from the stem, cut a hole in a piece of cardstock or thin cardboard large enough to fit the stem. The stem can then be put through the hole and the mushroom can be rested on top of a glass or jar.

Yellow Swamp Brittlegill

Russula claroflava

Look out for look-alikes

Of the more than 600 species, most Russula species are edible — but you have to be very confident in your identification. Smell is the one of the indicators for proper identification, as they each have distinctive odours. This one is fruity while others tend to be stinky or fishy.

Must-have features

1. Convex, flat or slightly concave round cap, depending on age.
2. Bright yellow to ochre-yellow cap that is smooth but sticky when wet. 4–10 cm across.
3. Densely packed, pale yellow gills.
4. Brittle white stem that snaps like chalk and darkens with age. 1–2 cm wide, 4–10 cm tall.
5. White flesh that stains grey with age.

Habitat
Grows in damp forests with birch and aspens.

SPORE COLOUR

BRUISING COLOUR

SMELL

Fruity

GROWTH SEASON

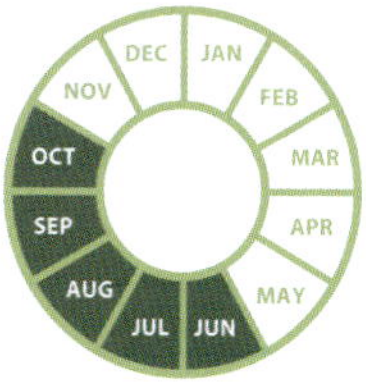

Late Oyster

Sarcomyxa serotina

Good to know

Sarcomyxa serotina appears in late fall, making it the last mushroom of the year. As a late season mushroom they are often bug-free in colder months.

Must-have features

1. Round, kidney-shaped or fan-shaped cap, depending on age. ✓
2. Olive green cap with violet or brown tones; fades to light brown with age. ✓
3. Cap is slightly sticky and slimy when wet. 3–10 cm across. ✓
4. Crowded yellowish gills become brownish with age. ✓
5. Slightly fuzzy stalk. 2–2.5 cm wide and long. ✓
6. Tough and rubbery white flesh. ✓

Habitat
Grows on decaying deciduous logs and stumps.

SPORE COLOUR

BRUISING COLOUR

No bruising change

SMELL

Indistinct

GROWTH SEASON

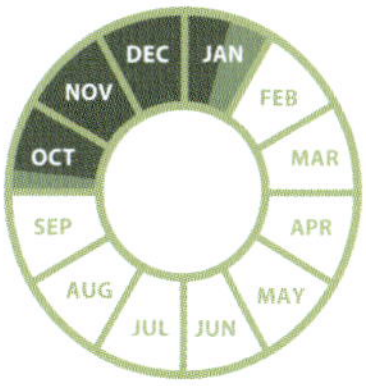

Slimy Spike

Gomphidius glutinosus

Look out for look-alikes

Other members of this genus that appear earlier on the East Coast are generally similar and are also edible. *Gomphidius borealis* and *Gomphidius maculatus* have more orange in the cap, while *Gomphidius nigricans* appears more reddish-pink before darkening with age.

Must-have features

1	Conical shape said to look like a bolt or child's spinning top.	✓
2	New growth is covered in slimy veil, which later disappears to reveal a grey-brown cap.	✓
3	Jelly-like purple, grey or brownish cap that may develop black spots with age. 4–9 cm across.	✓
4	Whitish gills extend down the stem and become grey with age.	✓
5	White stem with yellow at the base is widest at the bottom of the gills. 2.5–9.5 cm long, 1–2 cm wide.	✓
6	White flesh may be lemon yellow at the base.	✓

Habitat
Grows on the ground near conifers, usually spruce or fir.

SPORE COLOUR

BRUISING COLOUR

No bruising change

SMELL

Indistinct

GROWTH SEASON

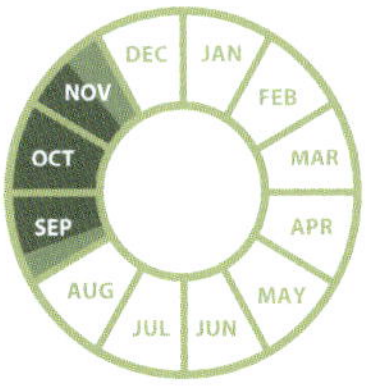

Blewit

Collybia nuda

Look out for look-alikes

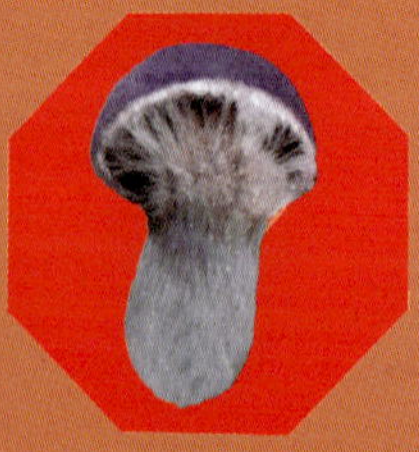

Do not eat this species unless you're familiar with the highly toxic webcap (*Cortinarius*) mushrooms. These are also purple and can live in similar locations but have a thin, weblike veil extending from the cap to the stalk. If brown spots are visible on the stem, it is not a blewit and should not be consumed.

Must-have features

1. Sticky lilac cap turns light brown with age. 3–12 cm across.
2. Crowded lilac gills are slightly stuck to stem.
3. Lilac stem is finely fibrous. 4–6 cm long, 2–2.5 cm thick.
4. DOES NOT have a veil on the stem.
5. Hygrophanous, meaning the colour of the cap fades with age or when drying out.

Habitat
Grows from plant debris near trees, such as clumps of leaves, conifer needles and sticks.

SPORE COLOUR

BRUISING COLOUR

No bruising change

SMELL

Fragrant and floral

GROWTH SEASON

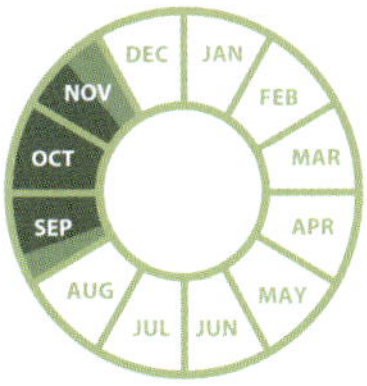

Purple and White Deceiver

Laccaria ochropurpurea

Look out for look-alikes

Make sure you are familiar with the toxic webcap mushrooms (*Cortinarius*) before eating this mushroom. These toxic look-alikes have a cobwebby veil over their gills during early development, and brown spores often collect on the remnants of the veil at maturity.

Must-have features

1. White cap with lilac tints may be slightly scaly. 3–12 cm across.
2. Waxy, dark purple gills attached to the stem. May become white with spores with age.
3. DOES NOT have a cobweb-like veil over gills.
4. Stem is off-white and fibrous, may discolour brownish. 5–19 cm tall, 1–2.5 cm wide.
5. Stem is more textured and disproportionately wide in comparison to the cap when young.

Habitat
Grows from the ground in deciduous forests, prefers to grow near oak and beech trees.

SPORE COLOUR

BRUISING COLOUR

SMELL

Indistinct

GROWTH SEASON

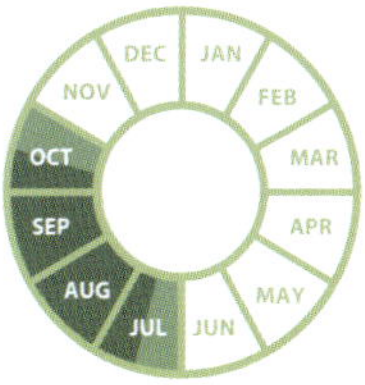

Indigo Milkcap

Lactarius indigo

Look out for look-alikes

Indigo Milkcaps are rare but pretty easily identified. The paler blue Anise Seed Funnel, *Clitocybe odora*, is a smaller look-alike, but it has a strong licorice odor and does not produce blue latex when cut.

Must-have features

1. Convex cap with a distinct impression in the middle that deepens over time. Mature specimens are almost funnel shaped, revealing the gills. 4–12 cm wide.
2. Vibrant indigo cap with occasional green tints fades to silvery-blue or grey over time. Oozes blue latex when cut or bruised. Cap is spotted and has a crater-like texture.
3. Closely spaced greenish-blue gills are attached to stem. Will stain fingers blue when bruised. Becomes dark green when exposed to the air.
4. Stalk tapers slightly at the base and can be off-centre relative to the cap. 2–8 cm long, 1.5–2 cm wide.
5. White flesh turns indigo blue when cut or bruised.

Habitat
Grows on the ground near coniferous and deciduous trees such as oaks, maples, pines and beeches. They are often buried under leaf litter.

SPORE COLOUR

BRUISING COLOUR

SMELL

Indistinct

GROWTH SEASON

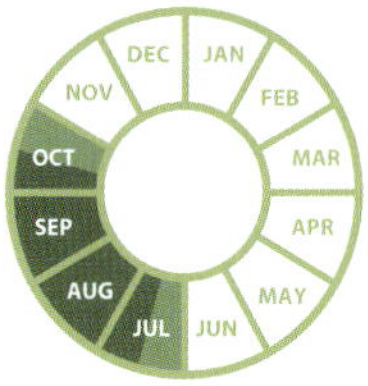

Brick Cap Mushroom

Hypholoma sublateritium

Look out for look-alikes

These are NOT beginner friendly mushrooms and have many toxic look-alikes. Watch out for *Leratiomyces ceres*, another orange-capped, purple-spored mushroom, which has a solid ring around its stem, tends to grow singly and has tufts on its cap. Other poisonous look-alikes include Sulphur tuft (*Hypholoma fasciculare*) and Funeral bell (*Galerina marginata*). Approach with extreme caution and eat only if identified by an expert.

Must-have features

1. Brick-red cap with white edges when young. 3–10 cm across.
2. Gills attach to stem, whitish when young, quickly turn purplish-grey.
3. Whitish stem with yellow at the top, brownish red at the bottom. 4–12 cm tall, 1–2 cm wide.
4. White cottony veil when young, leaves remnants near the top of the stem.
5. Firm whitish to yellowish flesh.

Habitat
Grows in clusters on decaying deciduous logs and stumps.

SPORE COLOUR

BRUISING COLOUR

SMELL

Indistinct

GROWTH SEASON

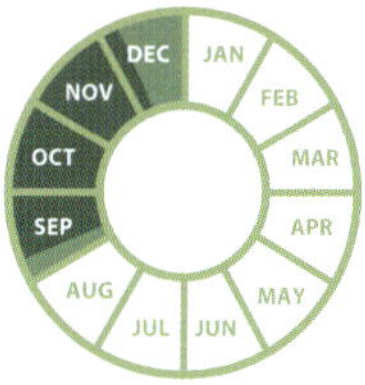

Field Mushroom

Agaricus campestris

Look out for look-alikes

There are close relatives, including *Agaricus xanthodermus,* which is yellow at the base of the stem and toxic. Always double-check that your stems are white.

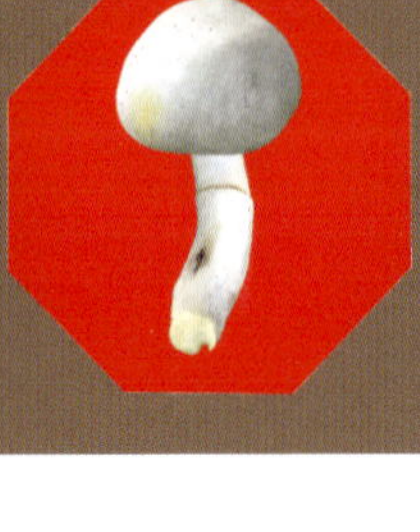

Must-have features

1. Pink when young, turning to brown as spores accumulate.
2. Creamy white cap is convex to flat. Appears smooth, dry or finely fibrous, and develops small scales with age, may become glossy. 4–10 cm wide.
3. Gills detached from stem. White veil covers gills when young, then lingers as a ring on the stem.
4. Stem is white and firm. 2–8 cm long, 1–2.5 cm wide.
5. White flesh.

Habitat
Grows from the ground on lawns and other grassy areas, sometimes along the edges of trails.

SPORE COLOUR

BRUISING COLOUR

SMELL

A slightly stronger version of the smell of white button mushrooms

GROWTH SEASON

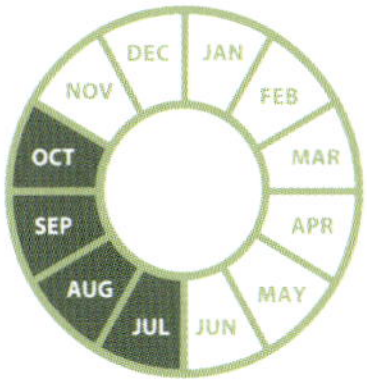

Dark Honey Mushroom

Armillaria ostoyae

Good to know

Eat only the cap and only in small quantities to avoid potential digestive issues.

Must-have features

1. Cap is light or dark brown with small brownish lump in the middle, covered in fine hairs. 5–15 cm across.
2. Gills are beige-white and attached to stem (occasionally extending down the stem).
3. Stalk is robust, tall, finger-width and white. At the base, stalk is slightly furry with brownish discoloration. 6–15 cm tall; 0.5–1.5 cm thick.
4. Well-defined white ring growing around the stalk below gills. This may fall off with age.
5. Stalk flesh is stringy when pulled apart.

Habitat
Grows in clusters on dead wood or on tree roots, and occasionally on tree trunks.

SPORE COLOUR

BRUISING COLOUR

No bruising change

SMELL

Usually indistinct, sometimes flour-like

GROWTH SEASON

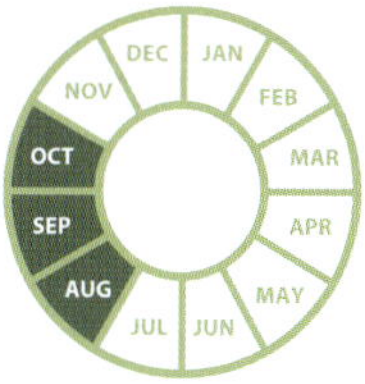

Aspen Oyster Mushroom

Pleurotus populinus

Look out for look-alikes

There are many oyster mushroom species, but this is the only one in the region found growing on aspen in June. Other edible species include *Pleurotus pulmonarius* and the *Pleurotus ostreatus* group, which grow in July or well into November.

Must-have features

1. Flat, shelf-like form.
2. Usually white, sometimes tan-brown, grey or grey-blue, becomes transparent to yellow with age.
3. Cap is mostly smooth, sometimes with fuzz near contact with tree. 4–17 cm across.
4. Gills are crowded and run down the stem to the tree.
5. Has a "pseudostem" , NOT a true stem.

Habitat
Grows in clusters on dead or dying aspen and poplar trees; usually on standing trees.

SPORE COLOUR

BRUISING COLOUR

No bruising change

SMELL

Licorice-like

GROWTH SEASON

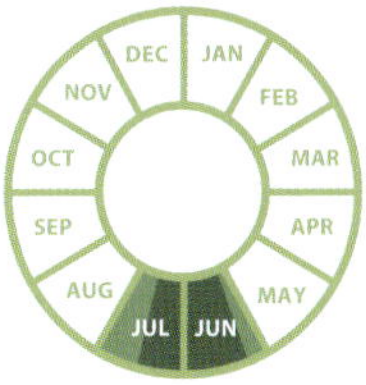

Shaggy Mane

Also known as Shaggy Inkcap or Lawyer's Wig

Coprinus comatus

Good to know

This mushroom's shelf life is very short (under twenty-four hours), since it liquifies as the mushroom breaks down. Do not eat this mushroom if the gills have started to turn inky.

Must-have features

1 Mushroom cap "melts" into a black spore-filled ink.

2 Cap is mainly white with thick fiber clumps darkening to beige-brown at the tips with age. Shaggy cap fibers and top discolour brown. 4–15 cm wide.

3 White, closely-packed gills appear separate from stem, turning black to slightly purple with age.

4 Stem is white and smooth with a cottony veil on lower or middle half. 1–2 cm wide, 5–20 cm long.

Habitat
Grows in disturbed ground including lawns, roadsides and woodchips.

SPORE COLOUR

BRUISING COLOUR

SMELL

Indistinct

GROWTH SEASON

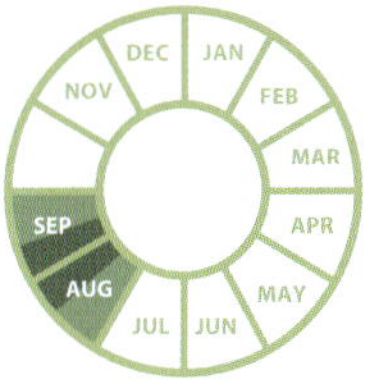

Larch Bolete

Also known as Tamarack Jack

Suillus grevillei

Look out for look-alikes

Similar species include the Ringless Slippery Jack, *Suillus granulatus.* As the name suggests, it has no stem ring, which is the key difference.

Must-have features

1. Shiny yellow cap with brown blotting around the edge darkens with age. Reddish caps sometimes occur. Yellowish pores on underside. 5–15 cm wide.
2. Greasy cap remains shiny even in dry conditions and feels slimy when wet. Flattens with age but retains a wavy shape.
3. Thin white translucent partial veil covers underside of young mushroom caps, forming a distinct white ring around the stalk just below the cap. Pale area remains after ring falls away.
4. Stalk is yellow above ring and streaked with red below. May widen slightly at base. 1–2 cm thick, 5–8 cm tall.
5. Yellow flesh has slight pink flush, does not change colour when cut.

Habitat
Common in grasslands and open forest areas on roots near Tamarack (Larch) trees.

SPORE COLOUR

BRUISING COLOUR

No bruising change

SMELL

Faint, but some describe as crumpled geranium leaves

GROWTH SEASON

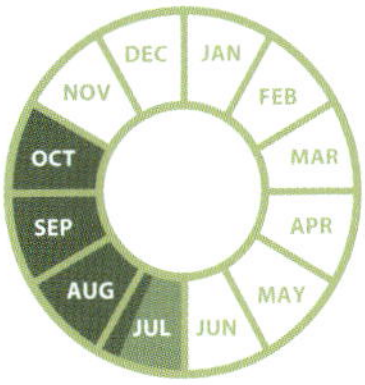

Smoothish-Stemmed Bolete

Hemileccinum subglabripes

Look out for look-alikes

While there may be boletes with yellow on them in the Maritimes, there is only one bolete around the Maritimes with yellow scabers on the yellow stem.

Must-have features

1. Cap is light brown, sometimes slightly wrinkled. 3–10 cm across.
2. Cap has yellow pores while fresh, which turn to olive in maturity.
3. Stem lined by fine yellow scab-like bumps, called scabers, which mature to reddish-brown from the bottom up.
4. Stalk sometimes curves at the end. 4–10 cm tall.
5. Flesh is yellow, and usually does not bruise blue. If it does, only slightly.

Habitat
Grows from the ground in deciduous forests.

SPORE COLOUR

BRUISING COLOUR

No bruising change

SMELL

Indistinct

GROWTH SEASON

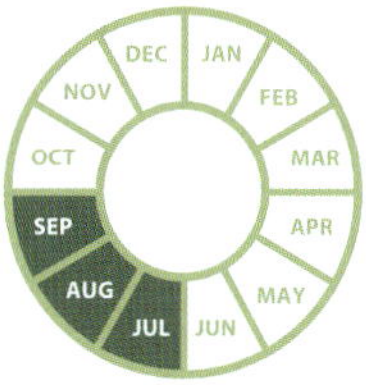

King Bolete

Also known as King Mushroom or Porcini

Boletus chippewaensis

Good to know

This mushroom is prized for its culinary versatility. May be listed as *Boletus edulis* in older field guides.

Must-have features

1. Cap is light brown, slightly wrinkled and often moist. 5–20 cm across.
2. Pores small, white at first, aging to yellow then olive.
3. Stem has a fishnet pattern and is often as wide as it is tall before it matures. 7–15 cm tall, 2–5 cm wide.
4. Flesh does NOT bruise blue.

Habitat
Grows on soil in mixed woods under conifers or deciduous trees.

SPORE COLOUR

BRUISING COLOUR

SMELL

Indistinct

GROWTH SEASON

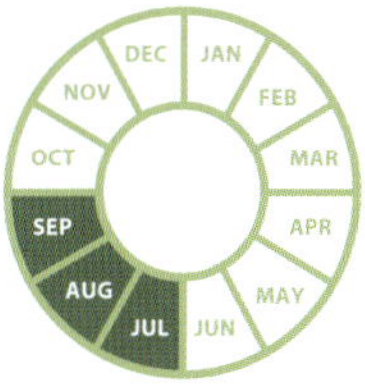

Bay Bolete

Imleria badia

Look out for look-alikes

When breaking or cutting these fungi open, your hands may stain dark blue. While blue bruising is often believed to indicate toxins, it's primarily a useful characteristic for identification. Some bluing mushrooms are edible, others aren't — do not use this to determine edibility.

Must-have features

1. Chestnut cap is large and smooth. Immature cap is softer and paler. 5–15 cm diameter.
2. Tubes and pores are pale yellow, turning blue-green when damaged.
3. Pale brown stalk with net-like pattern of vertical lines with brown ridges. Stem covered in fine hairs, slightly curved at base. 2–3 cm wide, 5–15 cm long.
4. Flesh of cap is whitish, occasionally light yellow, firm, tints blue when broken open.

Habitat
Typically grows in groups of 2–5 under mature trees.

SPORE COLOUR

BRUISING COLOUR

SMELL

Indistinct

GROWTH SEASON

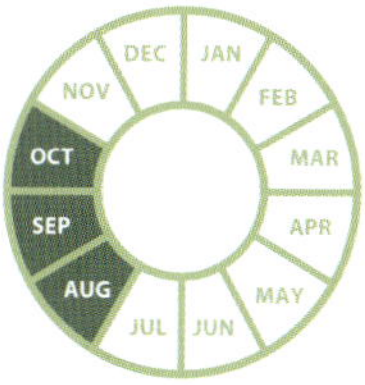

Dryad's Saddle

Cerioporus squamosus

Good to know

These mushrooms are edible when young, but as they age they become firmer, rubbery and possibly infested with maggots.

Must-have features

1. Saddle-shaped large brackets with large brown scales when growing on a tree trunk or log.
2. Funnel-shaped when found growing from soil.
3. Large tiered clusters growing on tree trunks 5–50 cm thick, 10–60 cm wide.
4. Stem often off-centre of cap; velvety and white. 8 cm long, 1–4 cm thick.
5. White flesh.

Habitat
Grows from the ground or on tree trunks or decaying logs up to several metres above the ground.

SPORE COLOUR

BRUISING COLOUR

No bruising change

SMELL

Like mealy potatoes or watermelon rind

GROWTH SEASON

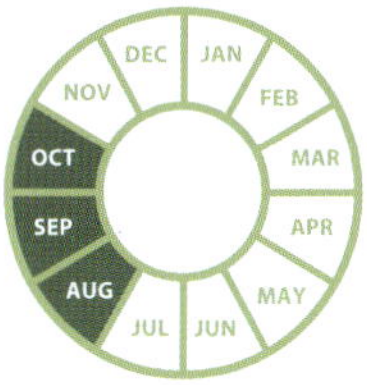

Brown Birch Bolete

Leccinum scabrum

Look out for look-alikes

Avoid orange-capped *Leccinum* mushrooms, as their close relative *Leccinum aurantiacum* is known to cause gastrointestinal distress.

Must-have features

1. Species may be sticky depending on conditions, and if cut, colour change is faint or absent.
2. Cap is varying shades of brown, dry, with an overhanging cap skin when young. 4–10 cm across.
3. Pores are white, aging yellowish, brown or grey.
4. Stem is fibrous, white, with rough black scab-like bumps, called scabers. Base may bruise blue. 6–12 cm tall, 1–1.5 cm across.
5. Cap flesh is soft and white, bruising grey over time.

Habitat
Grows in soil under birch trees.

SPORE COLOUR

BRUISING COLOUR

SMELL

Mild but indistinct

GROWTH SEASON

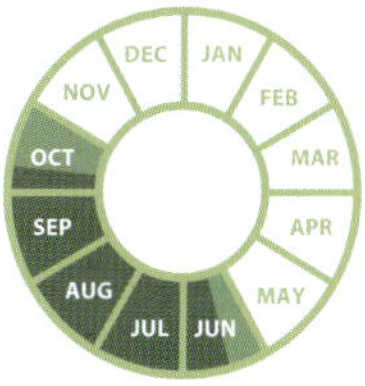

Chicken-of-the-Woods

Also known as Sulphur Polypore

Laetiporus sulphureus

Good to know

Larger specimens can survive overwintering. Aging mushroom flesh becomes tough and will decay into a black mush. The newest growth is found at the edge of the cap and can be cut away from the older flesh.

Must-have features

1. Cap has coloured layers ranging from egg-yolk yellow at tips to cream, pink and orange.
2. Cap is a wavy edged bracket without a stalk. 10–40 cm wide and 3–12 cm thick.
3. Tubes and pores are oval, white or pale yellow.
4. Pores DO NOT bruise black.
5. Flesh is yellow-orange, moist.

Habitat
Often found on standing tree trunks but also on decaying logs. Prefers deciduous species such as oak, beech, chestnut and cherry.

SPORE COLOUR

BRUISING COLOUR

No bruising change

SMELL

Slightly sweet but musty

GROWTH SEASON

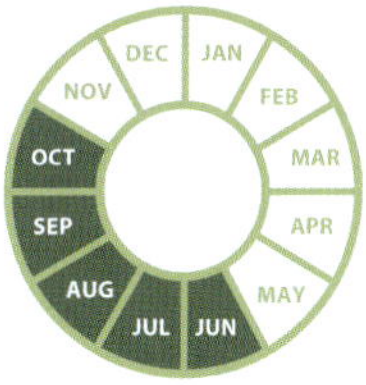

Bear's Toothed Fungus

Hericium americanum

Look out for look-alikes

Young fungus, before branches are visible, might be confused with *Hericium erinaceus* (which is also edible). This mushroom may also resemble *Hericium coralloides* which has longer branches and shorter spines.

Must-have features

1. Lacks gills or pores, instead being a "hydnoid" fungus with spines.
2. Off-white, 15–20 cm across.
3. Grows in clumps, with multiple toothed branchings per fruiting.
4. Flesh is soft, non-fibrous.

Habitat
Grows on dead or dying beech or oak trees.

SPORE COLOUR

BRUISING COLOUR

No bruising change

SMELL

Indistinct

GROWTH SEASON

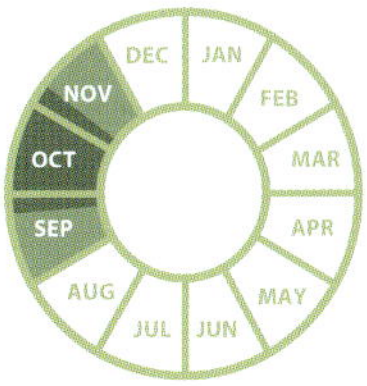

Toothed Jelly Fungus

Pseudohydnum gelatinosum

Good to know

There are no look-alikes this time! Being gelatinous, growing on wood and having spines rules out everything but this.

Must-have features

1. The whole fungus is translucent white to greyish, brown, rubbery and gelatinous.
2. Slightly convex, tongue- or kidney-shaped cap, gelatinous but not slimy, smooth to finely fuzzy on top. 2.5–6 cm across.
3. Fine spines along the bottom of the fungus, about 2 mm long.
4. Stem up to 3 cm long.
5. Flesh is gelatin-like, semi-translucent.

Habitat
Grows on decaying conifer wood, often hemlocks.

SPORE COLOUR

BRUISING COLOUR

No bruising change

SMELL

Indistinct

GROWTH SEASON

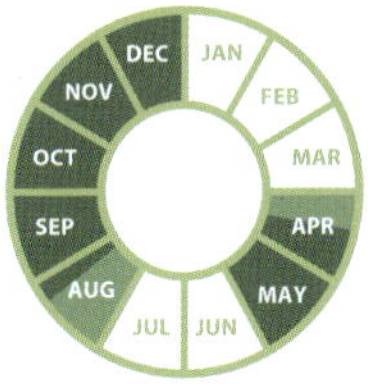

Wood Hedgehog

Also known as Sweet Tooth Mushroom

Hydnum repandum

Look out for look-alikes

A smaller but equally popular edible lookalike is the Depressed Hedgehog, *Hydnum umbilicatum,* which is named after its indented cap.

Must-have features

1. Tooth-like structures under the cap hang down like icicles. Soft brown colour with occasional hints of orange. 2–6 mm long.
2. Creamy white cap with fine velvety texture. 4–15 cm wide.
3. Irregularly shaped with wavy edges that may turn upward in more mature specimens.
4. Dry white or buff stalk. 6–10 cm tall, 2–3 cm wide.
5. Solid white flesh that breaks easily. Hints of orange staining when handled.
6. Grows in small clumps and occasionally in fairy circles.

Habitat
Grows on the ground in grassland and mixed forest, specifically near oaks, beeches, pines and spruces

SPORE COLOUR

BRUISING COLOUR

SMELL

Sweet but not distinct

GROWTH SEASON

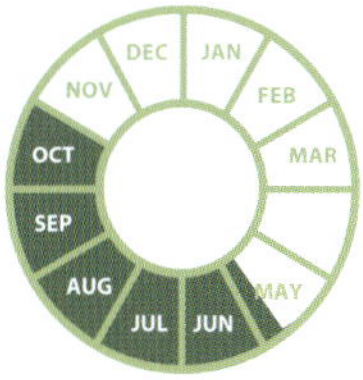

White Coral Fungus

Ramariopsis kunzei

Look out for look-alikes

The many varieties of coral fungi can be difficult to identify, and some may cause gastrointestinal distress. *Ramariopsis lentofragilis* is a toxic species that is easily confused with White coral fungus. Eat only if identified by an expert.

Must-have features

1. Small coral fungus with pure white branchings.
2. About 2–12 cm high, up to 10 cm wide.
3. Bluntly pointed branchings lack distinct "crown tips."
4. Flesh is usually brittle, but may be pliable.

Habitat
Grows from the ground in mixed woods or under deciduous trees.

SPORE COLOUR

BRUISING COLOUR

No bruising change

SMELL

Indistinct

GROWTH SEASON

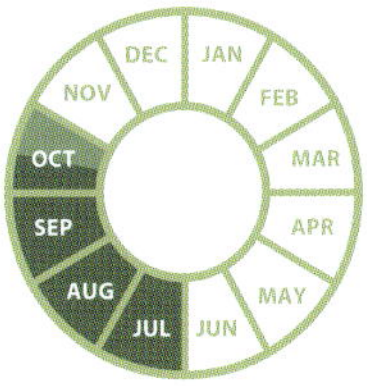

Crown Tipped Coral

Artomyces pyxidatus

Look out for look-alikes

The many varieties of coral fungi can be difficult to identify, and some may cause gastrointestinal distress. Eat only if identified by an expert.

Must-have features

1. Fruiting bodies are 4–15 cm high, 2–10 cm wide.
2. Fruiting bodies are white to translucent, branching like coral.
3. White or pale yellow growing from a short, greyish stalk-like base and branching out at top.
4. Branching ends have circular, crown-shaped cups adorned with 4–8 more points, which may be light brown at the tips.
5. Whitish, tough and pliable flesh.

Habitat
Grows on deciduous logs.

SPORE COLOUR

BRUISING COLOUR

No bruising change

SMELL

Smells faintly of freshly dug potatoes

GROWTH SEASON

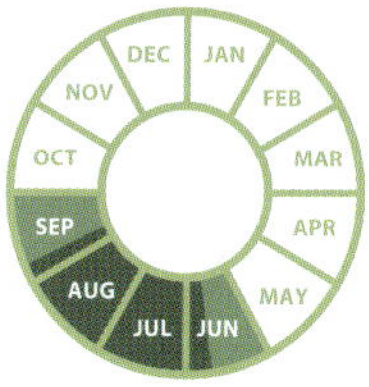

Golden Chanterelle

Cantharellus enelensis

Look out for look-alikes

The "False chanterelle" is toxic and must be avoided. It is more orange and has less-wrinkly gills than true chanterelles but can mislead beginners. The highly toxic *Cortinarius rubellus* can also be easily mistaken for a golden chanterelle and people have died from eating it. Older guides will refer to golden chanterelles as *Cantharellus cibarius.*

Must-have features

1. Golden yellow while young, can turn slightly orange with age.
2. Cap is finely velvety and firm. 3–11 cm diameter.
3. Not a true gilled mushroom. Gill-like wrinkles, called false gills, run down top portion of stem.
4. Stem is widest close to the cap. Can grow 4–12 cm high.
5. Flesh is white and stringy.

Habitat
Grows from the ground in sandy conifer habitats.

SPORE COLOUR

BRUISING COLOUR

SMELL

Fruity, apricot

GROWTH SEASON

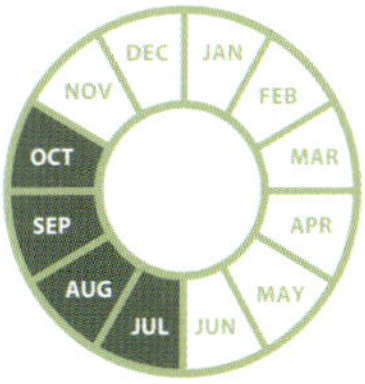

Winter Chanterelle

Craterellus tubaeformis

Good to know

These mushrooms are also called yellowfoot because of their yellow stalks, a feature that helps identify the species.

Must-have features

1. Cap is dark yellow to brownish grey, depressed in the middle. 2–7 cm wide.
2. Underside of cap lacks true gills, instead has grey ridges/folds that are considered false gills.
3. Yellow to orange stem. 3–9 cm long, under 1 cm thick.
4. Flesh ranges from yellowish to brownish.

Habitat
Grows near conifers, usually in moss near boggy areas.

SPORE COLOUR

BRUISING COLOUR

SMELL

Often indistinct, sometimes lightly fruity

GROWTH SEASON

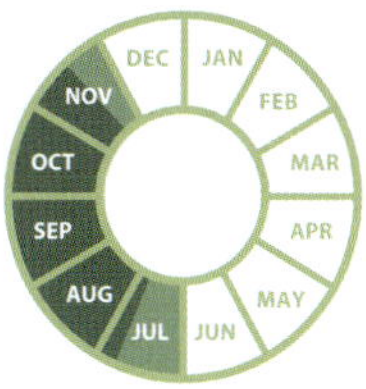

Cinnabar Chanterelle

Also known as Red Chanterelle

Cantharellus cinnabarinus

Good to know

A tip for differentiating between true and false gills: true gills will rub off from the mushroom, but false gills won't, since they are part of the flesh.

Must-have features

1. Bright brick red to orangey pink overall. Cap is 3–10 cm wide, convex but flattens and becomes wavy with age. Also pales in colour over time.
2. Has distinct false gills which look like thick, rounded wrinkles that fork and grow down portion of stalk.
3. Cinnabar red cap fades with age. 1.5–5 cm wide.
4. White or buff stalk that is dry and smooth to the touch. 4–6 cm tall, 1–2.5 cm wide.
5. Whiteish flesh with hints of the same bright colour as the cap. Does not change colour when exposed to air.

Habitat
Grows in clusters on the ground amongst hardwood trees such as oak and beech.

SPORE COLOUR

BRUISING COLOUR

No bruising change

SMELL

Fragrant, slightly sweet

GROWTH SEASON

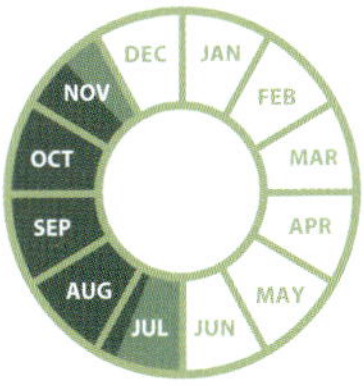

Black Trumpets

Craterellus fallax

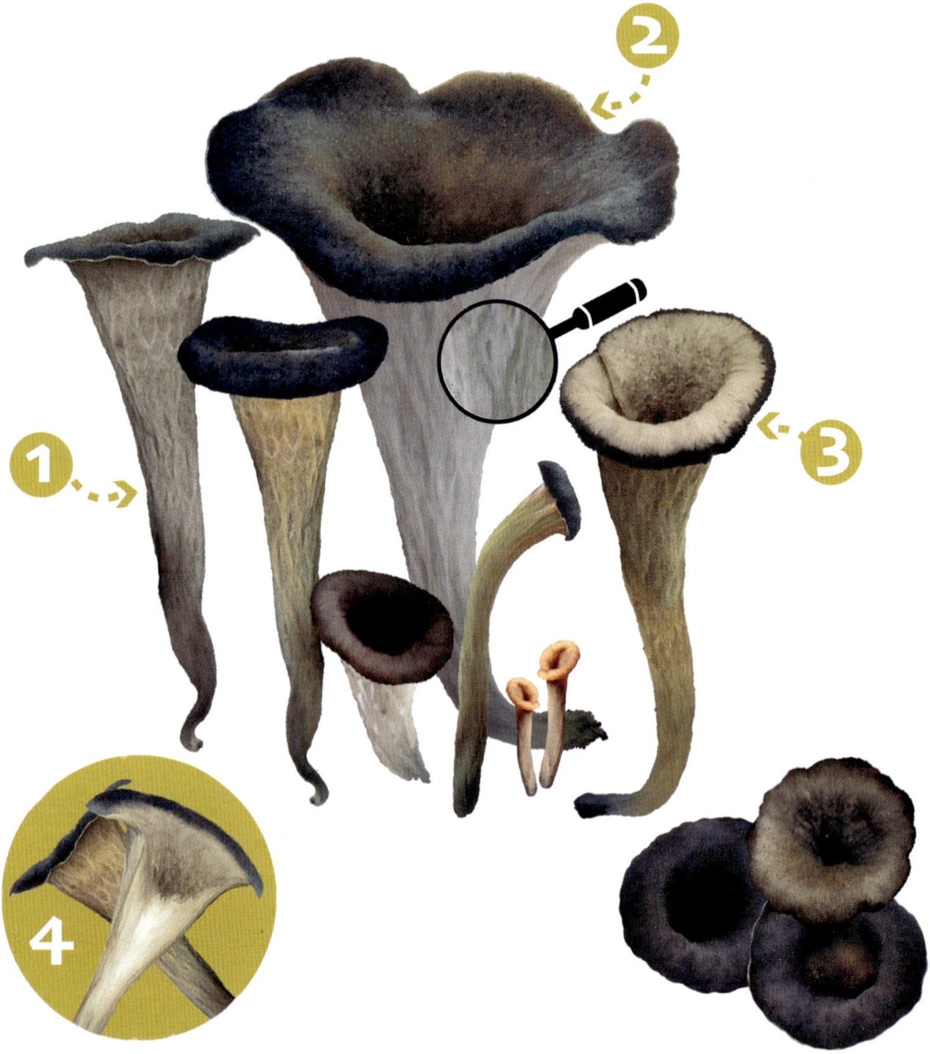

Good to know

Common and easily recognized, yet elusive to spot, the small black shape with a camouflaged body can be challenging to find. This mushroom has no toxic twins, making it one of the safer options for beginners.

Must-have features

1. Tubular shape at first, becoming a deep vase-shaped mushroom. Thin edges roll under.
2. No clearly defined cap or stem. 1–5 cm wide, 3–10 cm tall.
3. Grey to black on top surfaces and smoother, wrinkled outer surfaces. Skin develops yellow and orangish shades with maturity.
4. Thin and brittle grey to blackish flesh.

Habitat
Grows alone and scattered in loose clusters below oaks and other deciduous trees in mossy areas.

SPORE COLOUR

BRUISING COLOUR

No bruising change

SMELL

Indistinct

GROWTH SEASON

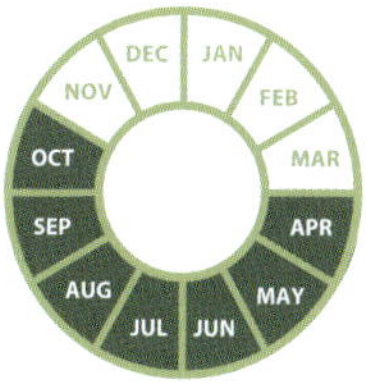

Lobster Mushroom

Hypomyces lactifluorum

Good to know

Lobster mushrooms are actually an edible parasitic mould that disfigures and digests *Russula* mushrooms, usually *Russula brevipes.*

Must-have features

1. Blob-like mushroom becomes chanterelle-like in shape with age. 6–20 cm across.
2. Somewhat hard and dotted with very tiny pimples.
3. Cap starts out white with orange specks and becomes solid orange over time.
4. Gills of host mushroom will be absent, or dissolve down to a fine ridge.
5. Little to no stem; fades to light colour at base.
6. Flesh is white, firm and brittle while fresh.

Habitat
Grows in soil under conifers, occasionally near deciduous trees.

SPORE COLOUR

BRUISING COLOUR

No bruising change

SMELL

Indistinct

GROWTH SEASON

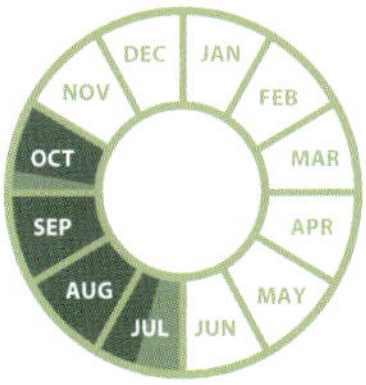

Shrimp of the Woods

Entoloma abortivum

Good to know

The mushroom has two forms: one is a typical mushroom form with a cap and a stalk, while the other is a blob-like shape. This second form is the result of *Entoloma abortivum* parasitizing a honey mushroom.

Must-have features

1. Popcorn-like specimens are off-white, or mottled grey and brown.
2. Specimens with more standard mushroom structure are grey to grey-brown. 2.8 cm across, 2–10 cm tall.
3. Stem is solid with enlarged base, bald or covered in fine hairs. 2–8 cm high, 5–15 mm thick.
4. Firm, white flesh, unchanging when cut open.

Habitat
Grows on or near dead or dying trees. Often seen near Honey mushrooms (page 36).

SPORE COLOUR

BRUISING COLOUR

SMELL

Mealy or watermelon rind-like

GROWTH SEASON

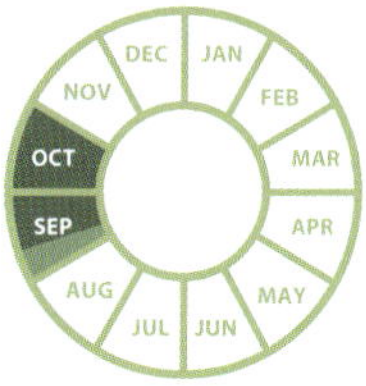

Common Puffball

Lycoperdon perlatum

Good to know

DO NOT breathe in the spores of a puffball. While it's tempting to step on them and shoot out spores, excessive inhalation can cause respiratory disease with symptoms including shortness of breath, muscle pains, fever and nausea. People with this condition have gone into comas, required breathing tubes and had portions of their lungs surgically removed.

Must-have features

1. Can go from slightly conical to almost spherical in shape. 3–8 cm tall, 2–7 cm wide
2. Skin has small, circular spines.
3. Skin is white while fresh, discolouring to brown with age.
4. Has a white or grey pseudo-stem that can either be shorter or longer than the main spherical portion.
5. Flesh is marshmallowy and white, which will become spores when mature.

Habitat
Grows on the ground, or on plant debris in conifer or deciduous forests, lawns or roadsides.

SPORE COLOUR

BRUISING COLOUR

No bruising change

SMELL

Indistinct

GROWTH SEASON

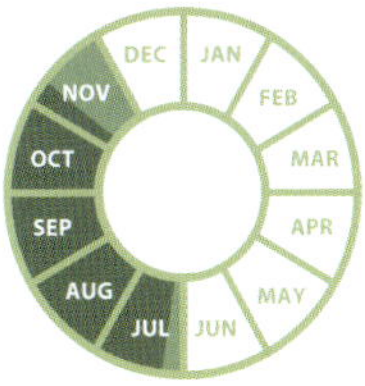

Giant Puffball

Calvatia gigantea

Good to know

Only harvest pure white mushrooms. Before bringing this mushroom home, cut it in half to confirm the specimen is pure white, free of any pests and does not have a stalk or gills. Harvest young specimens only that have smooth, uncracked skin.

Must-have features

1. Fully grown specimens can be as big and round as soccer balls or more blob-like. 20–30 cm across and 14–25 cm high on average. Can weigh multiple kilograms.
2. Outer skin, when fresh, is about 1 mm thick, pure white and has a soft velvet texture. As it ages, the skin turns yellow or olive brown, develops cracks and flakes.
3. No stalk, but has a short cord, 6 cm long and 2 cm thick, that grows out of the base and attaches it to the ground.
4. Flesh is firm and uniform white, resembling mozzarella cheese or tofu.

Habitat
Grows on the ground at the edge of forests, on lawns and in meadows. Grows in groups or alone.

SPORE COLOUR

BRUISING COLOUR

No bruising change

SMELL

Indistinct

GROWTH SEASON

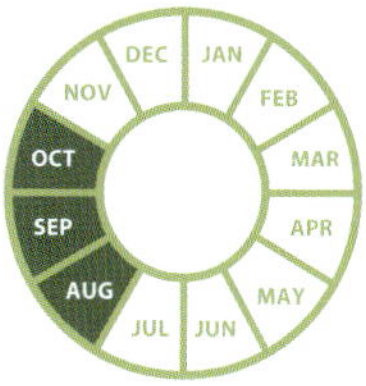

Cooking and Eating Wild Mushrooms

Yellow Swamp Brittlegill, 20
Russula claroflava

Pairs well with meat. Sauté in onion and garlic, add to an omelette or use in soups and stews. Has a pleasant, mildly nutty flavour with a hint of pepper.

Late Oyster, 22
Sarcomyxa serotina

Due to the sticky caps and potential compounds contained in the mushroom, it's advisable to cook this mushroom thoroughly. Taste varies from indistinct to slightly bitter, while the scent is indistinct.

Spike Cap, 24
Gomphidius glutinosus

Gelatinous portions should be removed before consuming. Smell and flavour are indistinct. Remove any slime along with skin of the cap immediately after picking. Recommended for cooking with other mushrooms in soups and stews. Not suited for drying.

Blewit, 26
Collybia nuda

Fries up excellently. Its taste has been described as sweet to slightly bitter, but pleasant. It may smell hay-like and pleasantly fragrant.

Purple and White Deceiver, 28
Laccaria ochropurpurea

Texture similar to the common button mushroom. Stick to using the cap of the mushroom, as the stem is fibrous and hard to digest. Unfortunately, the gills lose their vivid purple colour during the cooking process. Flavour is indistinct.

Indigo Milkcap, 30
Lactarius indigo

The texture is firm and somewhat crumbly when cooked. Their flavour is mild, though slightly bitter to some. Can be added to omelets, stir-fries and soups, or grilled. Their blue colour turns green with heat, so they can be used to make Green Eggs and Ham.

Brick Cap Mushroom, 32
Hypholoma sublateritium

Fry it, stew it, roast it. Taste slightly bitter, with a fairly faint, indistinct odour. Young mushrooms are insect free and edible when cooked. Refer to toxic cautions on species page.

Field Mushroom, 34
Agaricus campestris

Cook much like you would white button mushrooms found in grocery stores, as they're closely related and share similar flavours. If harvested at maturity, you may discover a richer taste.

Dark Honey Mushroom, 36
Armillaria ostoyae

No toxins, but difficult to digest. Discard older stems, as they can become fibrous. Pick young fungi for their sweet and earthy flavour, and consume within two days of harvesting. For thorough cooking, braise for at least fifteen to twenty minutes or boil before roasting. They are great additions to stews, soups, fillings or pasta sauces.

Aspen Oyster Mushroom, 38
Pleurotus populinus

Oyster mushrooms are often used in soups in Asia, but also fry up nicely when they're fresh and small. Their smell has been described as similar to aniseed or almond when fresh, and their flavour is sweet and pleasant.

Shaggy Mane (also known as Shaggy Inkcap or Lawyer's Wig), 40
Coprinus comatus

Fry or use in soups and sauces. Taste and smell indistinct.

Larch Bolete, 42
Suillus grevillei

Peel off slimy cap skin to avoid gastric irritation. Eat in small quantities at first to ensure you have no digestive reaction. Mild, earthy flavour and a texture that works in stews or stir-fries.

Smoothish-Stemmed Bolete, 44
Hemileccinum subglabripes

Fry or use in casseroles and chilis. Pick before the pores turn olive and mushy, but while the cap flesh remains firm. Stems edible on smaller specimens. Has a slight fruit taste with a distinct smoky, peppery flavour, stronger but less fruity than the golden chanterelle.

King Bolete (also known as King Mushroom or Porcini), 46
Boletus chippewaensis

One of the best mushrooms for the frying pan, although some people dry and powderize these mushrooms to thicken soups.

Bay Bolete, 48
Imleria badia

While considered a good edible mushroom, make sure to peel the cap skin off prior to cooking. Note that bay boletes contain xerocomic acid and are poisonous when eaten raw. The recommended cooking time is at least fifteen minutes.

Dryad's Saddle, 50

Cerioporus squamosus

After harvesting, scrape off pores with a fingernail to tell if the mushroom is tender. Avoid darker brackets and older specimens due to their toughness. Caps should be sliced into layers to check for insects and then cooked thoroughly.

Brown Birch Bolete, 52

Leccinum scabrum

Avoid using the stems as they're fibrous. On mature specimens, you will want to remove the pores, but on younger mushrooms you can cook the whole cap. Considered slightly toxic if consumed raw, so cook for a minimum of fifteen minutes. They are mild in taste and can be fried, or used in soups and sauces.

Chicken-of-the-Woods (also known as Sulphur Polypore), 54

Laetiporus sulphureus

Uncooked, it has a sour taste. Best harvested when young and moist. Cut into slices and pan-fry with oil, adding lemon for flavour. Taste is similar to chicken. Excellent for storage by freezing, as it retains its flavour well. NOTE: Toxins created if mushroom is growing on Ground Hemlock (Yew). Some harvesters have indicated feeling of nausea after eating.

Bear's Toothed Fungus, 56

Hericium americanum

One of the best mushrooms for frying. When cooked young, they are tender and meaty with a mild, nutty taste and a sweet fragrance of seafood, similar to lobster.

Toothed Jelly Fungus, 58

Pseudohydnum gelatinosum

Best saved for soups, since they're so watery they'll shrivel to nothing in a pan.

Wood Hedgehog, 60

Hydnum repandum

Harvest only young specimens, as they are less likely to contain insects or grubs. Their lookalikes, *Hydnum umbilicatum*, are also prized for their taste, but they are much smaller so a meal requires a larger harvest. These popular mushrooms have an earthy and slightly smoky flavour and are commonly eaten in stews and soups.

White Coral Fungus, 62

Ramariopsis kunzei

Fry or use in soup. This mushroom is quite versatile.

Crown Tipped Coral, 64

Artomyces pyxidatus

Its thin branchings crisp up nicely, so it's great for battering and frying. Tastes mildly peppery.

Golden Chanterelle, 66

Cantharellus enelensis

The stringy texture of this mushroom makes it a potential meat substitute. Its taste is often described as fruity, earthy, mildly peppery or reminiscent of apricot, while still retaining a distinct "mushroomy" flavour.

Winter Chanterelle, 68

Craterellus tubaeformis

Often simply fried as a standalone dish. They're small, so minimal preparation yields maximum flavour. Taste and smell may vary from indistinct to fruity.

Cinnabar Chanterelle, 70

Cantharellus cinnabarinus

Clean carefully with water as dirt can get stuck in their false gills. Sauté to remove the moisture and concentrate their peppery flavour before adding to omelets or pasta.

Black Trumpets, 72

Craterellus fallax

Considered one of the best-tasting mushrooms. It has been compared to truffle, with a fruity aroma and subtle apricot scent. The fruity aroma lends itself well to desserts.

Lobster Mushroom, 74

Hypomyces lactifluorum

Ensure you only pick specimens that are still firm and show minimal signs of bug damage. They remain recognizable long after they're still edible. Commonly added to creamy sauces and served with noodles. If it has begun to emit a fishy odour, it is no longer edible.

Shrimp of the Woods, 76

Entoloma abortivum

These make great stuffed mushrooms due to their plump shape. When browned in a pan they dry out and develop a delicious nutty flavour.

Common Puffball, 78

Lycoperdon perlatum

The size of this mushroom makes it versatile, but frying best highlights its meaty texture. Before consumption, always verify the mushroom's identity by slicing it open and checking for undifferentiated white flesh. If you observe anything other than undifferentiated white flesh, do not consume it. Mature puffball mushrooms are not edible. Once a "stink hole" forms at the top of the mushroom and emits a scent resembling carrion, DO NOT HARVEST.

Giant Puffball, 80

Calvatia gigantea

Wipe clean with soft damp cloth to remove dirt. Slice into mushroom "steaks" or cut into smaller pieces to grill, sauté or fry until golden-brown. Delicious with butter, garlic and white wine, or in pasta.

Acknowledgements

For their expertise, review of the artwork and careful fact-checking, we would like to thank the following people:

Alfredo Justo, the Curator of Botany and Mycology at the New Brunswick Museum. Justo is also a board member of the Société MycoNB Society, a member of the Mycological Society of America and associate editor of the journal *Mycologia.*

Dr. David Malloch, Emeritus Professor at the University of Toronto, Honorary Research Associate at UNB and creator of *Mycology Web Pages.*

Dr. Sean R. Haughian, Curator of Botany at the Nova Scotia Museum.

Index